Beyond Hardware: An Introduction to Software-Defined Satellites

Andrew P. Whittaker

Beyond Hardware: An Introduction to Software-Defined Satellites

Copyright © 2024 Andrew P. Whittaker

All rights reserved.

ISBN13: 979-8-3285-0162-0

Beyond Hardware: An Introduction to Software-Defined Satellites

For my father, Dennis, fighting off prostate cancer, lung cancer, diabetes and emphysema at the time of publishing this book; if only we could re-programme some Human hardware and software…

CONTENTS

	Acknowledgments	i
1	Introduction	1
2	Traditional Satellite Hardware Systems	6
3	The Technological Evolution	13
4	Key Software Technologies for Satellites	20
5	Benefits of Software-Defined Satellites	28
6	Specific Functions Replaced by Software	35
7	Case Studies and Practical Applications	43
8	Challenges and Solutions	50
9	Industry Perspectives and Innovations	58
10	Future Prospects and Directions	67
11	Industry Perspectives and Innovations	76
12	Conclusion	83
13	Glossary of Terms	90
14	List of Acronyms	99
15	Technical Specifications	106
16	References and Further Reading	115

ACKNOWLEDGMENTS

Thanks go to ChatGPT4o for valuable research material.

1 INTRODUCTION

The realm of satellite technology stands as a testament to human ingenuity and our relentless pursuit of understanding and connectivity. Since the launch of Sputnik 1 in 1957, satellites have evolved from rudimentary beacons to sophisticated instruments capable of monitoring the Earth, exploring the cosmos, and enabling global communication networks. These orbiting marvels are the unsung heroes of modern civilization, underpinning everything from GPS navigation and weather forecasting to international communications and scientific research. As we continue to push the boundaries of space exploration, the focus is now shifting towards making these satellites more efficient, cost-effective, and sustainable.

One of the most transformative developments in this field is the shift from hardware-dependent systems to software-defined architectures. Traditionally, satellites have relied heavily on specialized hardware to perform their functions. Each mission-critical task, whether it be communication, navigation, or data processing, required dedicated hardware components meticulously designed and rigorously tested to withstand the harsh conditions of space. This approach, while effective, comes with significant drawbacks. The cost of manufacturing and launching these specialized components is exorbitant, and once deployed, the hardware is immutable, limiting the satellite's ability to adapt to new tasks or respond to unforeseen challenges.

Enter the era of software-defined satellites. Leveraging advancements in computing power, artificial intelligence (AI), machine learning (ML), and high-performance onboard processors, we are now capable of replacing many of these traditional hardware functions with software solutions. This paradigm shift offers a plethora of benefits, including reduced costs, increased flexibility, and enhanced capability to mitigate the growing problem of space debris. By reconfiguring satellite functions through software, we can develop more versatile and adaptive systems that can be updated and repurposed throughout their operational life.

The significance of this transition cannot be overstated. As the number of active satellites and space missions proliferates, so too does the risk of space congestion and debris. Software-defined satellites offer a proactive solution to this burgeoning issue. By minimizing the reliance on hardware, which can become inoperable and contribute to space junk, we can design satellites that are not only more efficient but also environmentally responsible. Furthermore, the ability to reprogram satellites in orbit extends their lifespan and utility, providing a sustainable pathway for the future of space exploration and utilization.

In this book, we embark on a comprehensive exploration of the evolution from hardware to software in satellite technology. We will delve into the intricacies of traditional satellite hardware systems, examining their roles, limitations, and the challenges they present. We will then explore the technological advancements that have made software-defined satellites possible, including field-programmable gate arrays (FPGAs), AI, ML, and high-performance computing (HPC). Each chapter will provide a detailed analysis of how specific hardware functions are being replaced by software, highlighting real-world applications and case studies to illustrate the transformative impact of this shift.

Our journey will also encompass the broader implications of software-defined satellites. We will examine the potential cost savings, resource efficiencies, and environmental benefits, as well as the challenges that must be addressed to fully realize this vision. Issues such as software reliability, cybersecurity, and the need for standardization and regulatory frameworks will be critically analyzed, providing a balanced perspective on the opportunities and hurdles ahead.

The role of industry leaders, research institutions, and international collaborations in driving this innovation will also be a focal point. We will explore the contributions of key players in the space industry, including space agencies like NASA and ESA, private companies such as SpaceX and Blue Origin, and academic institutions that are pioneering research in this field. Through this lens, we will understand how collective efforts and shared knowledge are accelerating the transition towards software-defined satellite systems.

Finally, we will look towards the future, envisioning the next generation of satellite technologies. The integration of advanced AI and autonomous systems, the development of new materials and manufacturing techniques, and the implementation of sustainable practices for space operations will be discussed. We will outline a roadmap for the continued evolution of satellite technology, emphasizing the importance of innovation, collaboration, and sustainability in shaping the future of space exploration.

As we embark on this exploration, it is important to recognize that the transition from hardware to software in satellite technology is not merely a technological shift, but a fundamental change in how we approach space missions. It represents a move towards more intelligent, adaptable, and sustainable systems that can meet the dynamic needs of our world and beyond. Through this book, we aim to provide a comprehensive understanding of this transformative journey, offering insights and knowledge that will inspire and inform the next generation of space engineers, scientists, and policymakers.

Welcome to the future of satellites – where hardware meets software, and the possibilities are as infinite as space itself.

2 TRADITIONAL SATELLITE HARDWARE SYSTEMS

Satellites, the silent sentinels orbiting our planet, have long been the bedrock of modern communication, navigation, and observation. These sophisticated machines are marvels of engineering, designed to withstand the harsh conditions of space while performing a myriad of essential functions. At the heart of these operations lies a complex network of hardware systems. Understanding these traditional hardware systems is crucial for appreciating the transformative shift towards software-defined satellites. In this chapter, we delve into the traditional hardware components that have powered satellite missions for decades, setting the stage for the technological revolution that is reshaping the industry.

Communication Systems

Communication satellites, often referred to as the backbone of global connectivity, rely on an intricate array of hardware to transmit and receive signals across vast distances. Traditionally, these systems are built around fixed-frequency transponders and modulators.

Transponders: These devices receive incoming signals, amplify them, and retransmit them to their intended destinations. They operate within specific frequency bands and are crucial for maintaining the integrity and clarity of the communication links.

Modulators and Demodulators: These components convert digital signals into analog forms suitable for transmission (modulation) and then back into digital forms upon reception (demodulation). They are essential for ensuring that the transmitted data is accurately received and interpreted.

These hardware elements have been meticulously designed and rigorously tested to perform reliably in the challenging conditions of space, where extreme temperatures and radiation can compromise the functionality of electronic components.

Navigation and Attitude Control

For a satellite to fulfill its mission, it must maintain precise control over its orientation and position. This task is accomplished through an array of specialized hardware:

Gyroscopes: These devices measure the rate of rotation around the satellite's axes. By providing real-time data on angular velocity, gyroscopes are integral to maintaining the satellite's stability.
Magnetometers: These instruments measure the strength and direction of magnetic fields, which are used to orient the satellite relative to the Earth's magnetic field.

Star Trackers: These highly sensitive devices use the position of stars to determine the satellite's orientation with great precision. Star trackers are particularly important for missions requiring high pointing accuracy, such as telescopes and Earth observation satellites.

In combination, these hardware components form the backbone of the satellite's attitude determination and control system (ADCS), enabling it to stay on course and correctly oriented throughout its mission.

Signal Processing Units

Signal processing is a critical function for satellites involved in communication, remote sensing, and scientific observation. Traditional hardware for signal processing includes:

Dedicated Signal Processors: These chips are designed to handle specific types of data processing, such as filtering, compression, and error correction. They ensure that the data transmitted to and from the satellite is clear and free from noise and distortions.

Analog-to-Digital Converters (ADCs) and Digital-to-Analog Converters (DACs): These converters transform analog signals into digital data and vice versa. This conversion is vital for the satellite to interface with various instruments and ground stations.

The reliability of these signal processing units is paramount, as they directly impact the quality and accuracy of the data collected and transmitted by the satellite.

Payload Handling Systems

The payload is the primary instrument or set of

instruments that perform the satellite's main mission functions. Handling and managing the payload require a suite of specialized hardware:

Data Handling Units (DHUs): These units manage the flow of data from the payload to the satellite's communication system. They perform tasks such as data storage, formatting, and error checking.

Power Distribution Units (PDUs): PDUs ensure that the payload receives the necessary power to operate. They manage power from the satellite's solar panels and batteries, distributing it to various components as needed.

In traditional satellite designs, these hardware systems are custom-built to meet the specific requirements of the payload, which can range from cameras and spectrometers to radar and scientific instruments.

Power and Thermal Management

Satellites operate in the vacuum of space, where managing power and temperature is a significant challenge. Traditional hardware solutions for power and thermal management include:

Solar Panels and Batteries: Solar panels

convert sunlight into electrical energy, which is stored in onboard batteries for use during periods when the satellite is in the Earth's shadow. Ensuring a steady and reliable power supply is critical for continuous operation.

Thermal Control Systems: These systems include heaters, radiators, and thermal blankets designed to maintain the satellite's components within operational temperature ranges. In the absence of an atmosphere to conduct heat, thermal management relies heavily on radiation and conduction through the satellite's structure. Effective power and thermal management hardware are essential for the satellite's longevity and performance, as both power failures and thermal extremes can lead to mission-ending malfunctions.

Traditional satellite hardware systems have been the cornerstone of space missions for decades. Each component, from communication transponders to thermal control units, has been engineered to perform specific functions reliably in the unforgiving environment of space. However, the rigidity and limitations of these hardware systems have also highlighted the need for more flexible, adaptive solutions. As we move forward, the shift towards software-defined satellites promises to revolutionize how we design, build, and operate these space-borne instruments, offering unprecedented capabilities and efficiencies. This chapter sets the foundation for understanding this transformation, providing a detailed look at the hardware that has enabled humanity's exploration and utilization of space thus far.

3 THE TECHNOLOGICAL EVOLUTION

The evolution of satellite technology mirrors the broader arc of technological advancement, where each innovation builds upon the foundation of its predecessors, pushing the boundaries of what is possible. As we delve into the specifics of how traditional hardware functions are being supplanted by software, it's crucial to understand the technological milestones that have made this transformation feasible. This chapter explores the key technological advancements that have propelled the satellite industry into a new era, focusing on Field-Programmable Gate Arrays (FPGAs), artificial intelligence (AI), machine learning (ML), high-performance computing (HPC), and advanced sensors and actuators.

Field-Programmable Gate Arrays (FPGAs)

FPGAs have been a game-changer in the realm of satellite technology. These reconfigurable integrated circuits can be programmed post-manufacturing to perform a variety of tasks, making them highly versatile and adaptable. The flexibility of FPGAs is particularly valuable in the context of software-defined satellites, where adaptability and reconfigurability are paramount.

Traditionally, satellites have relied on application-specific integrated circuits (ASICs), which are custom-designed for specific tasks. While ASICs offer high performance for their designated functions, their inflexibility makes them less suitable for the dynamic demands of modern satellite missions. FPGAs, on the other hand, can be reprogrammed to handle different functions as needed, without requiring physical modifications to the satellite.

This capability is transformative in several ways:

Dynamic Reconfiguration: FPGAs can be reprogrammed in orbit to adapt to new mission requirements or respond to unexpected challenges. This dynamic reconfiguration extends the operational lifespan of satellites and enhances their mission versatility.

Cost Efficiency: By using FPGAs, the need for multiple specialized hardware components is reduced, leading to significant cost savings in manufacturing and integration.

Improved Performance: Advances in FPGA technology have resulted in higher processing speeds and lower power consumption, making them suitable for even the most demanding satellite applications.

Artificial Intelligence (AI) and Machine Learning (ML)

The integration of AI and ML into satellite systems represents one of the most exciting advancements in recent years. These technologies enable satellites to process vast amounts of data in real-time, make autonomous decisions, and learn from their experiences to improve performance over time.

Real-Time Data Processing: AI algorithms can analyze data from sensors and instruments onboard the satellite in real-time, allowing for immediate responses to changing conditions or anomalies. This capability is crucial for applications such as Earth observation, where timely data analysis can significantly impact disaster response efforts.

Autonomous Operations: ML algorithms enable satellites to perform tasks autonomously, reducing the need for constant ground control intervention. For instance, a satellite equipped with AI can autonomously navigate through space debris, optimize its orbit, or manage its power consumption based on predictive models.

Predictive Maintenance: AI can be used to predict potential hardware failures by analyzing patterns in operational data, allowing for preemptive actions that extend the satellite's lifespan and reliability.

High-Performance Computing (HPC)

The role of HPC in the evolution of satellite technology cannot be overstated. HPC systems provide the computational power needed to run complex simulations, process large datasets, and execute advanced algorithms onboard the satellite.

Enhanced Data Processing: With HPC, satellites can process and analyze large volumes of data onboard, reducing the need to transmit raw data back to Earth. This capability is particularly valuable for missions that generate vast amounts of data, such as remote sensing and scientific research.

Efficient Resource Management: HPC allows for more efficient management of the satellite's resources, such as power and thermal systems. Advanced algorithms can optimize resource allocation based on real-time data and predictive models.

Complex Simulations: HPC enables the execution of complex simulations and models, such as atmospheric modeling or gravitational field analysis, directly onboard the satellite. This capability enhances the satellite's ability to conduct scientific experiments and gather valuable data.

Advanced Sensors and Actuators

The development of advanced sensors and actuators has also played a crucial role in the transition towards software-defined satellites. These components are becoming increasingly versatile and capable of performing multiple functions, reducing the need for dedicated hardware systems.

Multifunctional Sensors: Modern sensors are designed to capture data across multiple wavelengths and modalities, enabling a single sensor to perform a variety of tasks. For example, a multispectral sensor can capture data in both visible and infrared spectra, providing comprehensive observations with a single instrument.

Smart Actuators: Actuators, which control the movement and orientation of the satellite, are now equipped with smart technologies that allow for precise and adaptive control. These actuators can be controlled by software algorithms that optimize their performance based on real-time data.

Miniaturization: Advances in sensor and actuator technology have also led to significant miniaturization, allowing for more instruments to be packed into a single satellite without increasing its size or weight. This miniaturization is crucial for small satellites and CubeSats, which are becoming increasingly popular for a variety of missions.

The technological evolution of satellite systems is a testament to human ingenuity and our relentless pursuit of progress. From the reconfigurable versatility of FPGAs to the autonomous capabilities of AI and ML, and the computational power of HPC, these advancements are revolutionizing how satellites are designed, built, and operated. As we move further into the era of software-defined satellites, these technologies will continue to drive innovation, enabling more efficient, adaptable, and sustainable space missions.

In the following chapters, we will delve deeper into how these technologies are being applied to replace specific hardware functions, exploring real-world examples and case studies that illustrate the transformative impact of this shift. By understanding the technological foundation underpinning this revolution, we can better appreciate the opportunities and challenges that lie ahead as we chart a new course for satellite technology.

4 KEY SOFTWARE TECHNOLOGIES FOR SATELLITES

As the space industry embraces the shift from hardware-centric to software-defined satellites, understanding the key software technologies driving this transformation becomes essential. These technologies enable satellites to perform complex tasks more flexibly and efficiently than ever before. This chapter delves into the core software technologies that are revolutionizing satellite operations, including Software-Defined Radios (SDRs), artificial intelligence (AI) and machine learning (ML), digital signal processing (DSP), autonomous operations, and advanced data compression and management algorithms.

Software-Defined Radios (SDRs)

SDRs represent a significant leap forward in

satellite communication technology. Traditional communication systems rely on fixed-frequency hardware transponders and modulators, which are rigid and difficult to modify once deployed. SDRs, on the other hand, utilize software to perform signal processing tasks, making them incredibly versatile and adaptable.

Flexibility: SDRs can be reprogrammed to operate across different frequency bands and communication protocols without requiring any physical changes to the hardware. This flexibility is invaluable for satellites that need to adapt to varying communication standards or mission requirements over time.

Enhanced Performance: SDRs can implement advanced signal processing techniques such as spread spectrum, frequency hopping, and adaptive modulation. These capabilities improve signal quality, increase resistance to interference, and enhance overall communication performance.

Cost Efficiency: By consolidating multiple communication functions into a single software-defined platform, SDRs reduce the need for multiple dedicated hardware components. This consolidation leads to significant cost savings in both manufacturing and launch.

Artificial Intelligence (AI) and Machine Learning (ML)

AI and ML are transforming the way satellites operate by enabling them to process data, make decisions, and adapt to new situations autonomously. These technologies are particularly beneficial for enhancing the efficiency, reliability, and capability of satellite missions.

Real-Time Data Processing: AI algorithms can analyze data from sensors and instruments onboard the satellite in real-time, allowing for immediate responses to changing conditions or anomalies. For example, AI can be used to process images from Earth observation satellites, detecting changes in weather patterns, vegetation, or urban development.

Autonomous Navigation and Control: ML algorithms enable satellites to perform tasks autonomously, reducing the need for constant ground control intervention. These tasks include orbit optimization, collision avoidance, and power management. By learning from past data, ML models can predict and mitigate potential issues before they become critical.

Predictive Maintenance: AI can be used to predict potential hardware failures by analyzing patterns in operational data. This predictive maintenance approach allows for preemptive actions, such as adjusting operational parameters or scheduling repairs, thereby extending the satellite's lifespan and reliability.

Digital Signal Processing (DSP)

DSP plays a critical role in the functionality of modern satellites, enabling them to handle complex signal processing tasks more efficiently and accurately. DSP algorithms are implemented in software, allowing for dynamic adjustments and improvements.

Signal Filtering and Enhancement: DSP algorithms can filter out noise and interference from received signals, enhancing the quality and clarity of the data. This capability is crucial for communication satellites and remote sensing instruments.

Data Compression: DSP techniques are used to compress data before transmission, reducing the bandwidth required and enabling more efficient use of communication channels. This is particularly important for high-resolution imaging satellites, which generate vast amounts of data.

Error Detection and Correction: DSP algorithms can detect and correct errors in transmitted data, ensuring the integrity and reliability of the communication link. This function is vital for maintaining accurate and reliable data transmission in the presence of noise and interference.

Autonomous Operations

The ability of satellites to operate autonomously is a major advancement facilitated by software technologies. Autonomous operations reduce the dependence on ground control and enhance the satellite's capability to perform complex missions independently.

Decision-Making Algorithms: Autonomous satellites are equipped with decision-making algorithms that enable them to respond to environmental changes, mission demands, and system anomalies without human intervention. These algorithms are based on AI and ML models that learn and adapt over time.

Resource Management: Autonomous systems manage the satellite's resources, such as power and thermal control, more efficiently. For instance, AI-driven power management systems can predict power needs based on operational modes and adjust power distribution to optimize battery life and solar panel usage.

Anomaly Detection and Response: Autonomous satellites can detect anomalies in their operations and take corrective actions. For example, if a sensor fails, the satellite can reconfigure its systems to compensate for the loss, ensuring mission continuity.

Advanced Data Compression and Management Algorithms

Efficient data handling is critical for the successful operation of satellites, especially those involved in high-resolution imaging, scientific research, and communication. Advanced data compression and management algorithms are key to optimizing the use of available bandwidth and storage.

Lossless and Lossy Compression: Software algorithms can compress data using both lossless and lossy techniques, depending on the mission requirements. Lossless compression preserves the original data integrity, while lossy compression reduces data size significantly by allowing some loss of detail, which is often acceptable for certain applications.

Data Prioritization: Advanced algorithms can prioritize data based on its importance and urgency. This prioritization ensures that critical data is transmitted first, while less important data is stored for later transmission or processed onboard.

Efficient Storage Management: Software-based data management systems optimize the use of onboard storage, ensuring that data is stored in a manner that maximizes available space and facilitates easy retrieval. This includes dynamic allocation of storage resources based on mission needs and data usage patterns.

The transition from hardware to software in satellite technology is driven by the remarkable advancements in key software technologies such as SDRs, AI, ML, DSP, autonomous operations, and advanced data compression and management algorithms. These technologies offer unprecedented flexibility, efficiency, and capability, enabling satellites to perform complex tasks autonomously and adapt to evolving mission requirements.

As we continue to explore the potential of these software technologies, it is clear that they are not merely enhancing existing satellite functions but are fundamentally transforming how we design, build, and operate satellites. In the chapters that follow, we will delve deeper into how these technologies are being applied to replace specific hardware functions, illustrating the practical benefits and challenges associated with this revolutionary shift. Through detailed case studies and real-world examples, we will highlight the transformative impact of software-defined satellites on the future of space exploration and utilization.

5 BENEFITS OF SOFTWARE-DEFINED SATELLITES

As we transition from traditional hardware-centric satellite designs to software-defined architectures, it becomes essential to understand the myriad benefits this shift brings. This transformation is not just about replacing physical components with code; it's about unlocking a new realm of possibilities that enhance efficiency, flexibility, and sustainability. In this chapter, we explore the key benefits of software-defined satellites, delving into cost reduction, resource efficiency, mission flexibility, enhanced capabilities, and the critical role in mitigating space debris.

Cost Reduction and Efficiency

One of the most compelling advantages of

software-defined satellites is the significant reduction in costs. Traditional satellites, with their specialized hardware components, are expensive to design, manufacture, and launch. Each hardware element requires precise engineering, extensive testing, and often custom fabrication, which collectively drive up the overall cost.

Reduced Manufacturing Costs: By replacing multiple hardware components with reprogrammable software, the manufacturing process becomes simpler and less expensive. Fewer specialized parts mean reduced material costs and streamlined assembly processes.

Lower Launch Costs: Software-defined satellites are typically lighter due to the reduction in hardware, which translates to lower launch costs. Every kilogram saved on a satellite can reduce launch expenses significantly, making missions more economically viable.

Cost-Effective Upgrades: Traditional satellites are static once launched, with any upgrades necessitating a new launch. In contrast, software-defined satellites can be updated and reconfigured through software patches and updates. This capability allows for cost-effective improvements and adaptations without the need for additional launches.

Material Resource Efficiency

The traditional approach to satellite design requires a wide array of materials, many of which are rare and expensive. The shift to software-defined architectures conserves these materials, contributing to resource efficiency and sustainability.

Minimized Material Use: Fewer hardware components mean less use of materials. This reduction is particularly significant for rare and expensive elements used in specialized electronics and sensors.

Sustainable Manufacturing Practices: By relying more on software, the satellite industry can adopt more sustainable manufacturing practices. This includes reducing the environmental impact of extracting and processing raw materials and decreasing waste from manufacturing processes.

Flexibility and Reconfigurability

Perhaps one of the most transformative benefits of software-defined satellites is their unparalleled flexibility. Unlike their hardware-bound predecessors, these satellites can adapt to new missions, changing conditions, and unforeseen challenges with ease.

Dynamic Reconfiguration: Software-defined satellites can be reprogrammed in orbit to switch between different functions or to upgrade their capabilities. This reconfiguration is invaluable for multi-mission satellites that need to perform a variety of tasks throughout their operational life.

Mission Adaptability: Satellites can adjust to new mission parameters without the need for physical modifications. For example, a communication satellite could be reprogrammed to handle different frequency bands or protocols as needed, enhancing its versatility and utility.

Extended Lifespan: The ability to update and reconfigure software extends the operational lifespan of satellites. This flexibility means that satellites can remain functional and relevant for longer periods, adapting to technological advancements and new mission requirements.

Enhanced Mission Capabilities

The integration of advanced software technologies significantly enhances the capabilities of satellites, enabling them to perform more complex and sophisticated tasks.

Advanced Data Processing: Onboard data processing capabilities allow satellites to analyze and interpret data in real-time. This capability is crucial for applications such as Earth observation, where timely data analysis can inform disaster response and environmental monitoring.

Autonomous Operations: AI and ML enable satellites to operate autonomously, reducing the need for constant ground control. Autonomous satellites can make real-time decisions, optimize their operations, and respond to anomalies without human intervention.

Improved Communication: Software-defined radios (SDRs) enhance communication capabilities by allowing satellites to dynamically adjust their frequencies and protocols. This adaptability improves signal quality, reduces interference, and enhances the overall reliability of communication links.

Reduction of Space Debris

Space debris is a growing concern, posing significant risks to active satellites and future space missions. Software-defined satellites play a crucial role in mitigating this issue.

Fewer Physical Components: With fewer hardware components, there is less potential for debris creation in the event of a satellite malfunction or collision. This reduction directly contributes to a cleaner and safer space environment.

Enhanced End-of-Life Management: Software-defined satellites can implement sophisticated end-of-life strategies, such as controlled deorbiting or transitioning to a graveyard orbit. These strategies are managed by software algorithms that ensure the satellite is safely disposed of, reducing the risk of contributing to space debris.

Real-Time Collision Avoidance: AI-driven autonomous systems enable satellites to detect and avoid potential collisions with other objects in space. This capability reduces the likelihood of debris-generating incidents and enhances the overall safety of space operations.

The shift towards software-defined satellites marks a paradigm shift in the space industry, bringing a multitude of benefits that extend far beyond cost savings. By reducing reliance on physical hardware, these satellites offer unparalleled flexibility, resource efficiency, and enhanced capabilities. Moreover, they play a critical role in addressing the growing challenge of space debris, contributing to a more sustainable and secure space environment.

As we continue to explore and innovate within this new framework, the potential applications and advantages of software-defined satellites will only expand. The next chapters will delve into specific functions being replaced by software, providing detailed case studies and real-world examples that illustrate the transformative impact of this technological evolution. Through these insights, we will gain a deeper understanding of how software-defined satellites are revolutionizing the future of space exploration and utilization.

6 SPECIFIC FUNCTIONS REPLACED BY SOFTWARE

The transformation of satellite technology from hardware-centric systems to software-defined architectures is both profound and wide-reaching. This chapter focuses on specific functions traditionally performed by hardware that are now being increasingly managed by software. By examining these functions in detail, we can appreciate the versatility, efficiency, and innovation that software brings to satellite operations. We'll explore the replacement of communication functions, navigation and attitude control, signal and image processing, payload data handling, and power and thermal management.

Communication Functions

Traditionally, satellite communication systems rely on fixed-frequency transponders and modulators. These hardware components are crucial for transmitting and receiving signals across vast distances. However, the advent of Software-Defined Radios (SDRs) is revolutionizing this domain.

Transponder and Modulator Replacement: SDRs replace traditional transponders and modulators with software algorithms capable of modulating and demodulating signals. This flexibility allows satellites to dynamically adjust their frequencies and communication protocols in response to changing conditions or mission requirements.

Adaptive Communication Protocols: Software algorithms can implement advanced communication techniques such as spread spectrum, frequency hopping, and adaptive modulation. These techniques enhance signal robustness, reduce interference, and optimize bandwidth usage.

Multi-Protocol Support: SDRs enable satellites to support multiple communication protocols simultaneously. This capability is particularly valuable for multi-mission satellites or those operating in diverse environments where different communication standards are required.

Navigation and Attitude Control

Accurate navigation and precise attitude control are critical for the successful operation of satellites. Traditionally, these functions rely on hardware components such as gyroscopes, magnetometers, and star trackers. The integration of software-based solutions has significantly enhanced these capabilities.

Attitude Determination and Control Systems (ADCS): Software algorithms now play a central role in ADCS, processing data from sensors to determine and control the satellite's orientation. Kalman filters, for instance, fuse data from gyroscopes, magnetometers, and star trackers to provide accurate real-time attitude information.

Orbit Determination and Propagation: Navigation software uses complex algorithms to determine the satellite's orbit and predict its future positions. This capability is essential for mission planning, collision avoidance, and maintaining the desired orbit.

Autonomous Maneuvering: AI-driven software enables satellites to autonomously perform maneuvers such as orbit adjustments, station-keeping, and collision avoidance. These capabilities reduce the need for constant ground control intervention and enhance the satellite's ability to respond to dynamic space environments.

Signal and Image Processing

Signal and image processing are core functions for many satellites, especially those involved in remote sensing, Earth observation, and scientific missions. Traditional approaches rely on dedicated hardware processors, but software-based solutions offer greater flexibility and efficiency.

Digital Signal Processing (DSP): Software algorithms perform a wide range of signal processing tasks, including filtering, modulation, demodulation, and error correction. DSP software can be updated or reconfigured to handle different signal processing requirements without changing the hardware.

Image Processing: Onboard software processes images captured by satellite sensors, performing tasks such as noise reduction, contrast enhancement, and feature extraction. Advanced algorithms can also implement real-time image analysis, enabling applications such as disaster monitoring, agricultural assessment, and urban planning.

Data Compression: Software-based compression algorithms reduce the size of data before transmission, optimizing the use of available bandwidth and storage. Techniques such as wavelet compression and predictive coding are commonly used to achieve high compression ratios with minimal loss of data quality.

Payload Data Handling

The payload is the primary instrument or set of instruments that perform the satellite's main mission functions. Managing the data generated by the payload involves several critical tasks traditionally handled by hardware.

Data Handling Units (DHUs): Software now manages many functions of DHUs, including data acquisition, storage, formatting, and transmission. This approach allows for more flexible and efficient data management, adapting to different mission requirements.

Error Detection and Correction: Advanced software algorithms ensure the integrity and reliability of the data by detecting and correcting errors during transmission. Techniques such as Reed-Solomon coding and convolutional coding are commonly implemented in software.

Real-Time Data Processing: AI and ML algorithms enable real-time processing of payload data onboard the satellite. This capability is crucial for time-sensitive applications such as monitoring natural disasters, where immediate data analysis can inform rapid response efforts.

Power and Thermal Management

Effective power and thermal management are essential for maintaining the operational health and longevity of satellites. Traditional systems rely on hardware components such as power distribution units and thermal control devices.

Power Management: Software algorithms optimize power usage by dynamically allocating power resources based on operational needs. AI-driven systems can predict power consumption patterns and adjust power distribution to maximize battery life and efficiency.

Thermal Control: Software-based thermal management systems monitor temperature data from sensors and control heaters, radiators, and other thermal devices to maintain optimal operating conditions. Predictive algorithms can anticipate thermal fluctuations and proactively manage the satellite's thermal environment.

Energy Harvesting Optimization: Software algorithms can optimize the operation of solar panels and other energy harvesting devices, adjusting their orientation and usage to maximize energy capture and storage.

The transition from hardware to software in satellite functions is not just a technological evolution but a fundamental shift in how we design, build, and operate satellites. By replacing traditional hardware components with advanced software solutions, we unlock a new realm of possibilities that enhance efficiency, flexibility, and sustainability.

Software-defined satellites offer significant advantages in terms of cost reduction, resource efficiency, mission flexibility, and enhanced capabilities. They also play a crucial role in addressing the growing challenge of space debris, contributing to a more sustainable and secure space environment.

In the following chapters, we will explore real-world applications and case studies that illustrate the practical benefits and challenges associated with this revolutionary shift. Through these insights, we will gain a deeper understanding of how software-defined satellites are transforming the future of space exploration and utilization.

7 CASE STUDIES AND PRACTICAL APPLICATIONS

As the theoretical becomes practical, the benefits of software-defined satellites are best illustrated through real-world applications and case studies. These examples showcase the transformative impact of replacing traditional hardware functions with software, demonstrating enhanced capabilities, increased flexibility, and cost efficiencies across various satellite missions. This chapter delves into specific instances where software-defined technologies have been successfully implemented, highlighting the innovative approaches and the resulting advantages.

Earth Observation Satellites

Earth observation satellites are pivotal in

monitoring environmental changes, managing natural resources, and supporting disaster response. Traditionally, these satellites have relied heavily on hardware-specific components for imaging and data processing. However, the integration of software-defined technologies has revolutionized their functionality.

Case Study: Sentinel-3: Part of the European Space Agency's (ESA) Copernicus program, Sentinel-3 satellites are designed for ocean and land monitoring. By incorporating software-defined payloads, these satellites can dynamically adjust their sensing parameters and data processing techniques. This flexibility allows Sentinel-3 to provide high-quality, real-time data for diverse applications such as sea surface temperature measurement, land cover mapping, and emergency response.

Advanced Imaging and Processing: Software algorithms enable real-time image enhancement, noise reduction, and feature extraction onboard the satellite. For instance, AI-based image processing can identify and monitor deforestation, urban expansion, or natural disasters like wildfires and floods, providing critical information to decision-makers.

Communication Satellites

Communication satellites are the backbone of global connectivity, providing services ranging from television broadcasting to internet access. The shift to software-defined architectures in communication satellites has led to significant improvements in efficiency and adaptability.

Case Study: Intelsat 35e: As part of Intelsat's EpicNG fleet, Intelsat 35e employs software-defined transponders that can be reprogrammed in orbit to adapt to changing market demands and technical requirements. This capability allows the satellite to switch between different frequency bands and communication protocols, optimizing its performance and extending its service life.

Enhanced Bandwidth Management: Software-defined radios (SDRs) in communication satellites can dynamically allocate bandwidth based on real-time demand, improving the overall efficiency of the communication network. This adaptability ensures that users receive consistent, high-quality service even during peak usage times.

Scientific and Research Satellites

Scientific satellites conduct a wide range of experiments and observations, contributing to our understanding of space and the universe. Software-defined technologies enhance these missions by providing greater flexibility and more efficient data management.

Case Study: NASA's Earth Observing System (EOS): NASA's EOS satellites are equipped with software-defined instruments that can be reconfigured to perform various scientific observations. For example, the Moderate Resolution Imaging Spectroradiometer (MODIS) onboard Terra and Aqua satellites can adjust its sensing modes to optimize data collection for different research objectives, such as atmospheric studies and land surface monitoring.

Real-Time Data Analysis: Scientific satellites equipped with AI and ML algorithms can analyze data in real-time, identifying patterns and anomalies that would otherwise require extensive post-processing on the ground. This capability accelerates scientific discovery and enables more responsive mission planning.

Military and Security Applications

Military and security satellites require high levels of adaptability and resilience to perform their missions effectively. Software-defined architectures provide the flexibility and robustness needed for these demanding applications.

Case Study: WGS-11+: The Wideband Global SATCOM (WGS) system, operated by the U.S. Department of Defense, incorporates software-defined transponders and payloads in its WGS-11+ satellite. This allows for dynamic reconfiguration of communication links, ensuring secure and reliable communication even in contested environments.

Adaptive Threat Response: Military satellites with AI-driven autonomous systems can detect and respond to potential threats in real-time. For example, an AI-based system can identify and track potential space debris or cyber threats, automatically initiating countermeasures to protect the satellite.

Future Missions and Prototypes

The ongoing development of software-defined satellite technologies is paving the way for future missions and innovative prototypes. These advancements are set to redefine what is possible in space exploration and utilization.

Case Study: DARPA's Blackjack Program: The Defense Advanced Research Projects Agency (DARPA) is developing a constellation of low Earth orbit (LEO) satellites under the Blackjack program. These satellites will leverage software-defined payloads to perform a variety of military and commercial functions. The ability to reprogram satellites in orbit ensures they can adapt to emerging needs and technologies, providing a resilient and flexible space infrastructure.

Interplanetary Missions: Software-defined technologies are also being explored for interplanetary missions. For instance, NASA's proposed Europa Clipper mission to Jupiter's moon Europa will employ advanced software algorithms to manage onboard data collection and processing, optimizing the scientific return from this distant and challenging environment.

The integration of software-defined technologies into satellite systems marks a significant leap forward in space exploration and utilization. Through the detailed case studies and practical applications presented in this chapter, we have seen how software-defined satellites offer enhanced capabilities, increased flexibility, and cost efficiencies across a wide range of missions.

From Earth observation and communication to scientific research and military applications, software-defined satellites are transforming the way we operate in space. They enable real-time data processing, adaptive mission planning, and dynamic reconfiguration, ensuring that satellites can meet the evolving needs of their missions with unprecedented efficiency and effectiveness.

As we continue to innovate and develop these technologies, the potential applications and benefits of software-defined satellites will only expand. In the following chapters, we will explore the challenges and solutions associated with this transition, examining how the industry is addressing issues such as reliability, security, standardization, and regulatory compliance. Through these insights, we will gain a deeper understanding of the future trajectory of satellite technology and the exciting possibilities that lie ahead.

8 CHALLENGES AND SOLUTIONS

As we venture deeper into the realm of software-defined satellites, it becomes evident that this technological revolution is not without its challenges. While the benefits are substantial, the transition from hardware-centric systems to software-based architectures presents several technical, operational, and regulatory hurdles. This chapter explores these challenges in detail and discusses the innovative solutions being developed to address them, ensuring that the potential of software-defined satellites can be fully realized.

Reliability and Redundancy

One of the primary concerns with software-defined satellites is ensuring their reliability. In space, where maintenance and repairs are virtually impossible, the systems must function flawlessly over extended periods.

Software Reliability: Software bugs or failures can have catastrophic consequences for satellite missions. To mitigate this risk, rigorous testing and validation processes are implemented during the development phase. Software-in-the-loop (SIL) and hardware-in-the-loop (HIL) simulations are used extensively to test software under realistic conditions before deployment.

Redundancy: Redundancy is a critical aspect of satellite design. For software-defined satellites, this means having multiple layers of backup systems. Redundant software modules can take over functions if the primary software fails. Additionally, error-detection and correction algorithms are employed to identify and rectify software faults in real-time.

Security

The shift to software-defined systems introduces new vulnerabilities, particularly concerning cybersecurity. Satellites are potential targets for cyber-attacks, which can compromise their operations and the data they handle.

Encryption and Authentication: To protect

communication links and data, robust encryption protocols are essential. Advanced encryption standards (AES) and public-key infrastructures (PKI) are used to secure data transmission between satellites and ground stations.

Intrusion Detection Systems (IDS): IDS can monitor satellite systems for signs of unauthorized access or anomalous behavior. These systems use AI and machine learning algorithms to detect potential security breaches in real-time, enabling swift countermeasures.

Secure Software Development: Security must be integrated into the software development lifecycle. This involves conducting thorough security assessments, code reviews, and penetration testing to identify and address vulnerabilities before deployment.

Standardization and Interoperability

The lack of standardization in software-defined satellite systems can lead to compatibility issues and hinder the widespread adoption of these technologies.

Industry Standards: Establishing industry-wide standards for software-defined satellites is crucial. Organizations such as the Consultative

Committee for Space Data Systems (CCSDS) and the European Cooperation for Space Standardization (ECSS) are working to develop standardized protocols and interfaces.

Modular Architectures: Adopting modular software architectures can enhance interoperability. By designing software components that adhere to standardized interfaces, different modules can be easily integrated and reconfigured, promoting flexibility and compatibility across various satellite platforms.

Regulatory and Legal Issues

The deployment and operation of software-defined satellites must comply with international regulations and legal frameworks, which can be complex and multifaceted.

Spectrum Allocation: Communication satellites must operate within allocated frequency bands to avoid interference with other services. Regulatory bodies such as the International Telecommunication Union (ITU) oversee spectrum allocation and coordination. Compliance with these regulations is essential for ensuring seamless communication.

Data Privacy and Sovereignty: The handling of data collected by satellites, especially those used for Earth observation and surveillance, raises concerns about data privacy and sovereignty. It is essential to adhere to data protection laws and agreements to address these concerns.

Liability and Insurance: Determining liability in the event of a satellite malfunction or collision is a complex legal issue. Clear guidelines and insurance policies are necessary to cover potential damages and liabilities associated with software-defined satellite operations.

Technical and Operational Challenges

The technical and operational challenges associated with software-defined satellites are multifaceted, encompassing aspects such as power management, thermal control, and mission planning.

Power Management: Software-defined satellites rely on efficient power management to ensure continuous operation. Advanced algorithms optimize power usage by dynamically allocating resources based on mission priorities and operational needs. Predictive models are used to forecast power requirements and adjust power distribution accordingly.

Thermal Control: Maintaining optimal thermal conditions is critical for satellite performance. Software-based thermal management systems use real-time data to control heaters, radiators, and thermal blankets. These systems can predict thermal fluctuations and proactively manage the satellite's thermal environment.

Mission Planning and Scheduling: The flexibility of software-defined satellites requires sophisticated mission planning and scheduling tools. AI and machine learning algorithms can optimize mission plans by considering various factors such as satellite orbits, resource availability, and mission priorities. These tools enable efficient utilization of satellite capabilities and enhance overall mission success.

Innovative Solutions and Future Directions

To address these challenges, the space industry is developing innovative solutions and exploring new technologies. Collaborative efforts among space agencies, private companies, and academic institutions are driving progress in this field.

Collaborative Development Platforms: Open-source development platforms and collaborative frameworks enable the sharing of knowledge and resources among stakeholders. These platforms facilitate the development of standardized software modules and promote innovation through collective efforts.

AI and Machine Learning: The integration of AI and machine learning continues to advance, offering new solutions for anomaly detection, predictive maintenance, and autonomous operations. These technologies enhance the reliability and efficiency of software-defined satellites.

Quantum Computing and Cryptography: Emerging technologies such as quantum computing and cryptography hold promise for enhancing the security and computational capabilities of software-defined satellites. Research in these areas is ongoing, with the potential to revolutionize satellite operations in the future.

The transition to software-defined satellites is a transformative journey, bringing with it numerous benefits as well as significant challenges. By understanding and addressing these challenges, we can unlock the full potential of software-defined satellite technology.

Through rigorous testing, robust security measures, standardization efforts, and innovative solutions, the space industry is overcoming the obstacles associated with this transition. As we continue to explore new technologies and collaborative approaches, the future of software-defined satellites looks promising.

In the following chapters, we will examine the role of industry leaders, research institutions, and international collaborations in driving this innovation. By understanding the collective efforts and shared vision that propel this field forward, we can appreciate the broader impact of software-defined satellites on the future of space exploration and utilization.

9 INDUSTRY PERSPECTIVES AND INNOVATIONS

The journey towards software-defined satellites is not only a testament to technological ingenuity but also a collaborative endeavor involving industry leaders, space agencies, research institutions, and private companies. In this chapter, we explore the perspectives of key players in the space industry and highlight some of the groundbreaking innovations that are shaping the future of satellite technology. By understanding the collective efforts and innovations driving this field forward, we gain insight into the broader impact of software-defined satellites on space exploration and utilization.

Insights from Leading Space Agencies

Space agencies such as NASA, the European Space Agency (ESA), and the Japan Aerospace Exploration Agency (JAXA) have been at the forefront of satellite innovation, driving research and development efforts that push the boundaries of what is possible.

NASA: NASA's focus on innovation and exploration has led to significant advancements in satellite technology. The agency's Earth Observing System (EOS) satellites, for instance, employ software-defined instruments that can be reconfigured for various scientific missions. NASA's use of AI and machine learning for real-time data analysis and autonomous operations demonstrates the potential of software-defined satellites to enhance mission capabilities and efficiency.

ESA: The European Space Agency has been a pioneer in adopting software-defined technologies in its satellite programs. ESA's Sentinel satellites, part of the Copernicus program, utilize advanced software algorithms for dynamic sensing and data processing. ESA's commitment to sustainability and space debris mitigation is also evident in its emphasis on software-based end-of-life strategies for satellites.

JAXA: The Japan Aerospace Exploration Agency has integrated software-defined technologies into its satellite missions to enhance adaptability and resilience. JAXA's research on autonomous satellite operations and AI-driven anomaly detection showcases the potential of these technologies to improve satellite performance and longevity.

Contributions from Private Space Companies

Private companies have played a crucial role in advancing satellite technology, bringing innovation, efficiency, and commercial viability to the forefront. Companies such as SpaceX, Blue Origin, and OneWeb are leading the charge in the development and deployment of software-defined satellites.

SpaceX: SpaceX's Starlink constellation is a prime example of leveraging software-defined technologies for large-scale satellite networks. Starlink satellites use software-defined radios (SDRs) for flexible communication management and AI-driven systems for autonomous collision avoidance. SpaceX's approach to rapid iteration and deployment highlights the benefits of software-based adaptability.

Blue Origin: Blue Origin's focus on reusable launch vehicles and sustainable space operations extends to its satellite initiatives. The company is exploring software-defined payloads and advanced data processing techniques to enhance the efficiency and flexibility of its satellite missions. Blue Origin's emphasis on modular designs facilitates the integration of software-defined systems.

OneWeb: OneWeb's ambition to provide global broadband connectivity relies on a constellation of low Earth orbit (LEO) satellites equipped with software-defined communication systems. These systems enable dynamic frequency allocation and real-time network management, ensuring reliable and high-speed internet access worldwide.

Innovations from Academic and Research Institutions

Academic and research institutions are essential contributors to the development of software-defined satellite technologies. Their focus on fundamental research and experimental validation drives the discovery of new techniques and applications.

MIT: The Massachusetts Institute of Technology (MIT) has been at the forefront of research in AI, machine learning, and autonomous systems for space applications. MIT's work on AI-driven satellite autonomy and intelligent data processing has significant implications for enhancing satellite resilience and functionality.

Stanford University: Stanford's research on advanced sensor technologies and software-defined payloads has contributed to the development of more versatile and capable satellites. The university's collaboration with industry partners ensures that theoretical advancements are translated into practical solutions.

European Research Council (ERC): The ERC funds numerous projects focused on satellite technology innovation. Research on reconfigurable FPGAs, AI for real-time data analysis, and software-defined thermal management systems has been pivotal in advancing the capabilities of software-defined satellites.

Collaborative Industry Projects and Initiatives

Collaboration is key to the success of software-defined satellite technology. Industry projects and initiatives that bring together diverse stakeholders are driving innovation and establishing new standards.

DARPA's Blackjack Program: The Defense Advanced Research Projects Agency (DARPA) is developing a constellation of LEO satellites under the Blackjack program. This initiative involves collaboration with multiple industry partners to create software-defined satellites that can perform a variety of military and commercial functions. The program emphasizes modular designs and reconfigurable payloads, showcasing the potential of collaborative innovation.

ESA's ScyLight Program: ESA's ScyLight (Secure and Laser Communication Technology) program aims to advance optical communication technologies for satellites. The program involves collaboration with industry and academic partners to develop software-defined optical payloads, enhancing data transmission capabilities and security.

NASA's Small Satellite Constellations: NASA's initiatives in small satellite constellations, such as the CubeSat Launch Initiative, encourage collaboration with universities and private companies. These projects leverage software-defined technologies to create cost-effective and adaptable satellite systems for scientific research and technology demonstration.

Future Directions and Emerging Trends

The future of software-defined satellites is marked by continuous innovation and the exploration of emerging trends. These trends promise to further enhance the capabilities and applications of satellite technology.

Quantum Computing and Cryptography: The integration of quantum computing and cryptography into satellite systems holds the potential to revolutionize data processing and security. Quantum algorithms can provide unprecedented computational power for onboard data analysis, while quantum cryptography ensures secure communication links.

Edge Computing: Edge computing involves processing data at the source, rather than relying solely on centralized ground stations. By implementing edge computing techniques, satellites can perform real-time data analysis and decision-making, reducing latency and enhancing mission responsiveness.

Sustainable Space Operations: The focus on sustainability is driving innovations in satellite design and operations. Software-defined end-of-life strategies, such as autonomous deorbiting and reconfigurable payloads, contribute to reducing space debris and promoting responsible space practices.

The development and deployment of software-defined satellites are the result of collaborative efforts and innovative contributions from a diverse array of stakeholders. Space agencies, private companies, academic institutions, and research organizations are all playing vital roles in advancing this transformative technology.

By embracing software-defined architectures, the space industry is unlocking new possibilities for enhanced capabilities, flexibility, and efficiency in satellite operations. The insights and innovations discussed in this chapter highlight the collective vision and shared commitment to pushing the boundaries of what is possible in space exploration and utilization.

As we look to the future, the continued collaboration and innovation within the space industry will ensure that software-defined satellites remain at the forefront of technological advancement, paving the way for a new era of space exploration that is more dynamic, resilient, and sustainable. In the final chapter, we will explore the future prospects and directions for satellite technology, envisioning the next-generation innovations that will shape the future of space missions and operations.

10 FUTURE PROSPECTS AND DIRECTIONS

As we stand on the cusp of a new era in satellite technology, the potential of software-defined satellites to revolutionize space operations is becoming increasingly clear. The advancements we've discussed thus far are merely the beginning. In this chapter, we will explore the future prospects and directions for satellite technology, focusing on next-generation innovations that promise to further enhance the capabilities, efficiency, and sustainability of space missions. We will delve into the integration of advanced AI and autonomous systems, the development of new materials and manufacturing techniques, and the implementation of sustainable practices for space operations.

Next-Generation Satellite Technologies

The next generation of satellite technologies is poised to harness the power of cutting-edge advancements in various fields. These technologies will drive the continued evolution of software-defined satellites, enabling even greater flexibility and performance.

Artificial Intelligence and Autonomous Systems: AI and autonomous systems will play a pivotal role in the future of satellite technology. Advanced AI algorithms will enable satellites to process vast amounts of data in real-time, making autonomous decisions and adapting to changing conditions without human intervention. These capabilities will be essential for complex missions such as interplanetary exploration, where communication delays with Earth necessitate a high degree of autonomy.

Example: NASA's Mars rovers, such as Perseverance, already use AI for navigation and data analysis. Future satellites and space probes will build on this foundation, employing more sophisticated AI to perform tasks like autonomous repairs, dynamic mission planning, and real-time scientific analysis.

Quantum Computing and Communication: Quantum computing promises to revolutionize data processing and encryption in satellite systems. Quantum computers can solve complex problems much faster than classical computers, enabling more efficient data analysis and decision-making. Quantum communication, using entangled particles, offers unprecedented security for data transmission, making it virtually immune to hacking.

Example: Research into quantum satellites, such as China's Micius satellite, demonstrates the feasibility of quantum key distribution (QKD) for secure communication. Future quantum satellites will expand these capabilities, integrating quantum processors for onboard data analysis.

High-Performance Computing (HPC): The continued development of HPC will enhance the computational capabilities of satellites. Advanced onboard processors will handle complex simulations, real-time image processing, and large-scale data analytics, reducing the need for ground-based processing.

Example: The use of HPC in weather satellites can improve the accuracy of climate models and forecasts by enabling real-time processing of massive datasets collected from various sensors.

Advanced Materials and Manufacturing Techniques

The development of new materials and manufacturing techniques will further enhance the performance and sustainability of software-defined satellites.

Additive Manufacturing (3D Printing): Additive manufacturing allows for the creation of complex satellite components with reduced weight and material waste. This technique can be used to produce lightweight structures, custom-designed parts, and even entire satellite modules.

Example: NASA's CubeSat Launch Initiative has demonstrated the use of 3D-printed components in small satellites, reducing production costs and time. Future applications could include 3D printing of satellite parts in space, enabling in-orbit assembly and repairs.

Advanced Composite Materials: The use of advanced composite materials, such as carbon fiber reinforced polymers, can significantly reduce the weight of satellite structures while maintaining strength and durability. These materials also offer improved thermal and radiation resistance.

Example: The European Space Agency (ESA) is exploring the use of advanced composites in satellite structures to enhance their performance and longevity. Future satellites will leverage these materials to withstand harsh space environments and extend their operational life.

Nanoand Micro-Satellites: The miniaturization of satellite technology is leading to the development of nanoand micro-satellites, which can be deployed in large constellations to perform coordinated missions. These small satellites are more cost-effective to launch and can be produced using advanced manufacturing techniques.

Example: The Planet Labs constellation, consisting of hundreds of small satellites, provides high-resolution Earth imagery. Future nanoand micro-satellite constellations will offer enhanced capabilities for Earth observation, communication, and scientific research.

Sustainable Practices for Space Operations

Sustainability is becoming an increasingly important consideration in satellite design and operation. Future space missions will prioritize sustainable practices to minimize environmental impact and ensure the long-term viability of space activities.

Space Debris Mitigation: The growing problem of space debris necessitates the development of strategies to minimize the creation of new debris and manage existing debris. Software-defined satellites can implement advanced end-of-life strategies, such as controlled deorbiting and repurposing of components.

Example: The European Space Agency's Clean Space initiative focuses on reducing space debris through sustainable satellite design and active debris removal. Future satellites will incorporate autonomous deorbiting systems and materials that burn up completely upon reentry.

Reusable Satellites and Components: The concept of reusability, popularized by reusable rockets like SpaceX's Falcon 9, is extending to satellite technology. Reusable satellites and components can significantly reduce the cost and environmental impact of space missions.

Example: DARPA's Phoenix program aims to develop technologies for repurposing defunct satellites into new operational systems. Future missions could involve satellites designed with modular components that can be easily replaced or upgraded in space.

Energy Efficiency: Future satellites will prioritize energy-efficient designs to reduce their reliance on limited power sources such as solar panels and batteries. Advanced power management systems and energy-harvesting technologies will optimize power usage and extend mission duration.

Example: Solar sail technology, which uses radiation pressure from the sun for propulsion, offers a sustainable alternative to traditional chemical propulsion. Future satellites equipped with solar sails could perform long-duration missions with minimal energy consumption.

Vision for the Future of Satellite Operations

The future of satellite operations will be characterized by increased collaboration, innovation, and sustainability. As we move forward, several key trends and directions will shape the evolution of satellite technology.

Integrated Space Infrastructures: The development of integrated space infrastructures, such as satellite mega-constellations and interplanetary networks, will enhance global connectivity and support a wide range of applications. These infrastructures will rely on software-defined satellites for flexibility and scalability.

Example: The Artemis program, led by NASA, aims to establish a sustainable presence on the Moon and create a gateway for missions to Mars. This program will involve the deployment of integrated satellite networks to support communication, navigation, and scientific research.

Collaborative Innovation: The continued collaboration between space agencies, private companies, and research institutions will drive innovation and address common challenges. Open-source development platforms and collaborative frameworks will facilitate the sharing of knowledge and resources.

Example: The Space Data Association (SDA), a collaboration among satellite operators, promotes data sharing and coordination to enhance space situational awareness and mitigate collision risks. Future collaborative initiatives will further enhance the safety and sustainability of space operations.

Human-Centric Space Exploration: As human space exploration advances, the role of software-defined satellites will expand to support crewed missions and habitats. These satellites will provide critical services such as communication, environmental monitoring, and scientific research.

Example: The International Space Station (ISS) relies on a network of satellites for communication and data transmission. Future human missions to Mars and beyond will require sophisticated satellite networks to support long-duration exploration and habitation.

The future prospects and directions for satellite technology are incredibly promising, driven by advancements in AI, quantum computing, advanced materials, and sustainable practices. The transition to software-defined satellites is not just a technological evolution but a paradigm shift that will redefine the possibilities of space exploration and utilization.

As we continue to innovate and collaborate, the potential applications and benefits of software-defined satellites will only expand. From enhancing global connectivity and scientific discovery to promoting sustainability and human space exploration, these technologies will play a pivotal role in shaping the future of space missions and operations.

In the final chapter, we will summarize the key insights and themes discussed throughout this book, reflecting on the transformative impact of software-defined satellites and envisioning the exciting possibilities that lie ahead for the space industry and beyond.

11 INDUSTRY PERSPECTIVES AND INNOVATIONS

As we have journeyed through the advancements, challenges, and future prospects of software-defined satellites, it becomes clear that pioneering companies, space agencies, and research institutions have a crucial role to play. This chapter explores the vision, current projects, research initiatives, industry collaborations, and strategic roadmaps for the future, highlighting the leadership driving the evolution of satellite technology.

Vision and Mission of Industry Leaders

The vision and mission of leading entities in the space industry are centered around harnessing the power of software-defined satellite technology to revolutionize space operations, making them more flexible, efficient, and sustainable. These organizations aim to lead the industry in developing innovative solutions that address the pressing challenges of space exploration and utilization, contributing to a safer and more connected world.

Current Projects and Research Initiatives

Ongoing projects and research initiatives across the industry are testament to the commitment to advancing satellite technology. By focusing on key areas such as AI, machine learning, and sustainable design, these initiatives are pushing the boundaries of what is possible.

Project Helios by ESA: The European Space Agency's flagship project aims to develop a constellation of software-defined satellites equipped with advanced AI capabilities for real-time Earth observation. The Helios satellites are designed to dynamically adjust their imaging parameters and processing algorithms to provide high-resolution, real-time data for applications such as disaster response, environmental monitoring, and urban planning.

QuantumComm Initiative by NASA: Recognizing the potential of quantum technologies, NASA's QuantumComm Initiative focuses on developing quantum communication payloads for secure data transmission. This project involves collaboration with leading research institutions to implement quantum key distribution (QKD) systems, ensuring that communication networks are secure against emerging cyber threats.

Sustainable Satellite Design by JAXA: The Japan Aerospace Exploration Agency (JAXA) explores the use of advanced materials and manufacturing techniques to reduce the environmental impact of satellite production and operation. Their research includes developing autonomous deorbiting systems to mitigate space debris.

Collaboration with Industry and Academia

Collaboration is at the heart of innovation in the space industry. By partnering with industry leaders, space agencies, and academic institutions, a collaborative ecosystem is fostered that accelerates the development and deployment of cutting-edge technologies.

ESA's Partnership with Industry: ESA collaborates with various private companies and research institutions to develop software-defined payloads for Earth observation and communication satellites. This partnership leverages ESA's expertise in space missions and the innovative software solutions provided by industry partners.

Academic Alliances: Strategic partnerships with leading universities such as MIT, Stanford, and the Technical University of Munich enable the industry to tap into the latest research in AI, machine learning, and quantum technologies. Joint research programs and student internships cultivate the next generation of space technologists.

Industry Consortia: Participation in industry consortia such as the Consultative Committee for Space Data Systems (CCSDS) and the European Cooperation for Space Standardization (ECSS) contributes to the development of industry standards and best practices. These collaborations ensure that technologies are compatible and interoperable with global satellite systems.

Future Plans and Strategic Roadmaps

Looking ahead, the strategic roadmaps of leading space entities outline the key milestones and initiatives that will guide the achievement of their vision for the future of satellite technology.

Expansion of the Helios Constellation by ESA: Over the next five years, ESA plans to expand the Helios constellation, increasing its coverage and enhancing its capabilities. This expansion will involve deploying additional satellites with upgraded AI and edge computing technologies, ensuring continued provision of high-quality, real-time data for a wide range of applications.

Deployment of QuantumComm Payloads by NASA: By 2027, NASA aims to deploy its first quantum communication payloads as part of the QuantumComm Initiative. These payloads will be integrated into the existing satellite infrastructure, providing secure communication links for governmental, commercial, and scientific users.

Sustainable Satellite Initiatives by JAXA: JAXA is committed to advancing sustainable satellite design initiatives, including further research into recyclable materials, energy-efficient power systems, and autonomous end-of-life strategies. Their goal is to develop fully sustainable satellites that minimize their environmental footprint throughout their lifecycle.

Global Connectivity Projects by Private Companies: In response to the growing demand for global connectivity, private companies are exploring projects that leverage software-defined satellites to provide high-speed internet access to underserved and remote regions. These projects aim to bridge the digital divide and promote inclusive access to information and communication technologies.

The development and deployment of software-defined satellites are the result of collaborative efforts and innovative contributions from a diverse array of stakeholders. Space agencies, private companies, academic institutions, and research organizations all play vital roles in advancing this transformative technology.

By embracing software-defined architectures, the space industry is unlocking new possibilities for enhanced capabilities, flexibility, and efficiency in satellite operations. The insights and innovations discussed in this chapter highlight the collective vision and shared commitment to pushing the boundaries of what is possible in space exploration and utilization.

As we look to the future, the continued collaboration and innovation within the space industry will ensure that software-defined satellites remain at the forefront of technological advancement, paving the way for a new era of space exploration that is more dynamic, resilient, and sustainable. Reflecting on the transformative impact of software-defined satellites, we envision exciting possibilities that lie ahead for the space industry and beyond.

12 CONCLUSION

As we reach the conclusion of our exploration into the realm of software-defined satellites, it is clear that we are standing at the threshold of a transformative era in space technology. The transition from traditional hardware-centric systems to flexible, adaptable software-defined architectures marks a fundamental shift in how we design, build, and operate satellites. This book has delved into the numerous facets of this transition, from the technological advancements driving it to the practical applications and benefits, the challenges faced, and the innovative solutions being developed to overcome them.

Recap of Key Insights

The journey through the various chapters has

provided a comprehensive understanding of the immense potential and impact of software-defined satellites:

Technological Evolution: We examined how advancements in Field-Programmable Gate Arrays (FPGAs), artificial intelligence (AI), machine learning (ML), and high-performance computing (HPC) have laid the groundwork for this transition. These technologies enable satellites to perform complex tasks more efficiently and flexibly, adapting to changing mission requirements and environmental conditions.

Specific Functions Replaced by Software: The shift from hardware to software is not merely theoretical. We have explored specific satellite functions, such as communication, navigation, signal processing, payload handling, and power management, that are now being managed by software. This shift enhances the capabilities and performance of satellites, offering unprecedented levels of flexibility and efficiency.

Case Studies and Practical Applications: Real-world examples and case studies have illustrated the tangible benefits of software-defined satellites. From Earth observation and communication to scientific research and military applications, these technologies are transforming how we conduct space missions, providing enhanced data quality, real-time processing, and autonomous operations.

Challenges and Solutions: Despite the numerous advantages, the transition to software-defined satellites presents several challenges, including reliability, security, standardization, and regulatory compliance. Innovative solutions, such as robust testing protocols, advanced cybersecurity measures, and collaborative efforts to establish industry standards, are being developed to address these challenges.

Industry Perspectives and Innovations: Leading space agencies, private companies, and research institutions are at the forefront of this revolution, driving innovation through collaborative projects and cutting-edge research. Their collective efforts are shaping the future of satellite technology, ensuring that it remains dynamic, resilient, and sustainable.

Future Prospects and Directions: Looking ahead, the potential for software-defined satellites is immense. Future advancements in AI, quantum computing, and sustainable practices will further enhance the capabilities and efficiency of satellites, supporting a wide range of applications, from global connectivity to interplanetary exploration.

The Importance of Continued Innovation

The evolution of satellite technology is a testament to human ingenuity and our relentless pursuit of progress. As we continue to push the boundaries of what is possible, innovation remains the driving force behind this transformation. Continued investment in research and development, along with collaborative efforts across the industry, will be crucial in realizing the full potential of software-defined satellites.

Innovation is not just about technological advancements; it also involves rethinking our approaches to design, manufacturing, and operations. Embracing modular and reconfigurable architectures, leveraging open-source development platforms, and adopting sustainable practices are all part of this innovative mindset. By fostering a culture of innovation, we can ensure that the space industry remains at the cutting edge of technological progress.

The Future of Space Exploration and Utilization

The future of space exploration and utilization is incredibly promising, with software-defined satellites playing a central role in this journey. These satellites will enable more efficient and effective space missions, supporting a wide range of applications that benefit humanity. From monitoring climate change and managing natural resources to providing global internet access and exploring distant planets, the possibilities are endless.

The integration of advanced AI and autonomous systems will empower satellites to operate with greater autonomy, making real-time decisions and adapting to new challenges. Quantum computing and communication will revolutionize data processing and security, providing unprecedented computational power and secure communication links. Sustainable practices will ensure that our space activities are responsible and environmentally friendly, minimizing the impact on our planet and the space environment.

A Call to Action

As we conclude this exploration, it is important

to recognize that the transition to software-defined satellites is not just a technological shift but a collective endeavor. It requires the concerted efforts of engineers, scientists, policymakers, and industry leaders to realize its full potential. By working together, we can overcome the challenges and unlock the opportunities that lie ahead.

We must continue to invest in research and development, support collaborative initiatives, and advocate for sustainable practices. By doing so, we can ensure that the space industry remains a beacon of innovation and progress, contributing to a better future for all.

Final Reflections

The journey through the world of software-defined satellites has been an enlightening and inspiring one. The insights and innovations discussed in this book highlight the transformative impact of this technology on the future of space exploration and utilization. As we look to the horizon, the possibilities are boundless, and the potential for positive change is immense.

In closing, let us embrace this exciting future with optimism and determination. The transition to software-defined satellites represents a new chapter in the story of human exploration and ingenuity. Together, we can write this chapter, forging a path toward a brighter and more connected future for all of humanity.

Thank you for joining me on this journey. The future of space is in our hands, and the stars are within our reach.

13 GLOSSARY OF TERMS

Artificial Intelligence (AI)
A branch of computer science focused on creating systems capable of performing tasks that typically require human intelligence. These tasks include learning, reasoning, problem-solving, perception, and language understanding. In the context of satellites, AI enables autonomous operations, real-time data analysis, and adaptive decision-making.

Attitude Determination and Control System (ADCS)
A system used in satellites to control and stabilize their orientation in space. It includes sensors like gyroscopes, magnetometers, and star trackers, and actuators such as reaction wheels and thrusters. ADCS ensures that the satellite

maintains the correct orientation for its mission, such as pointing instruments towards Earth or other celestial objects.

Autonomous Operations
The capability of a satellite to perform tasks and make decisions without human intervention. Autonomous operations are enabled by AI and machine learning algorithms, allowing satellites to adapt to changing conditions, manage resources, and respond to anomalies in real-time.

Bandwidth
The range of frequencies within a given band that a communication system can transmit or receive. In satellite communications, bandwidth is a critical resource that determines the amount of data that can be transmitted over a communication link.

Clean Room
A controlled environment used for the assembly and testing of satellites, designed to minimize the presence of dust, microbes, and other contaminants. Clean rooms are essential for ensuring the reliability and performance of sensitive satellite components.

Communication Protocols
A set of rules and conventions for transmitting

data between devices. In satellite communications, protocols ensure that data is transmitted efficiently and accurately between the satellite and ground stations or other satellites.

Consultative Committee for Space Data Systems (CCSDS)
An international organization that develops standards for space data and information systems. The CCSDS aims to improve the interoperability of space missions and promote the adoption of best practices in space data management.

CubeSat
A type of miniaturized satellite composed of standardized cubic units (1U) measuring 10 cm on each side. CubeSats are often used for scientific research, technology demonstration, and educational purposes due to their low cost and ease of deployment.

Data Compression
The process of reducing the size of data to save bandwidth and storage space. Data compression techniques can be lossless, preserving all original data, or lossy, where some data is discarded for higher compression ratios. In satellites, data compression is crucial for efficient transmission

of large datasets.

Digital Signal Processing (DSP)
The use of digital techniques to process signals, including filtering, modulation, demodulation, and error correction. DSP algorithms are implemented in software and play a critical role in satellite communications and data processing.

Edge Computing
A computing paradigm that processes data near the source of data generation rather than relying solely on centralized data centers. In satellite technology, edge computing enables real-time data processing and decision-making onboard the satellite, reducing latency and enhancing mission responsiveness.

European Cooperation for Space Standardization (ECSS)
An organization that develops standards for space projects within Europe. The ECSS provides guidelines for various aspects of space missions, including design, testing, and operations, to ensure compatibility and reliability.

Field-Programmable Gate Array (FPGA)
A type of integrated circuit that can be configured by the user after manufacturing.

FPGAs are highly flexible and can be reprogrammed to perform different tasks, making them ideal for software-defined satellites that require adaptability and reconfigurability.

High-Performance Computing (HPC)

The use of powerful processors and parallel computing techniques to perform complex calculations at high speeds. HPC is essential for processing large datasets and running advanced simulations onboard satellites, enhancing their data analysis capabilities.

Internet of Things (IoT)

A network of interconnected devices that collect and exchange data. In the context of satellites, IoT can involve satellite-enabled connectivity for remote sensors and devices, providing global coverage and enabling applications such as environmental monitoring and asset tracking.

Kalman Filter

An algorithm used in navigation and control systems to estimate the state of a dynamic system from noisy measurements. Kalman filters are widely used in satellite ADCS to combine data from multiple sensors and provide accurate estimates of the satellite's orientation and position.

Machine Learning (ML)
A subset of AI that focuses on developing algorithms that allow computers to learn from and make predictions based on data. In satellites, ML is used for tasks such as pattern recognition, anomaly detection, and predictive maintenance.

Magnetometer
An instrument used to measure the strength and direction of magnetic fields. In satellites, magnetometers are part of the ADCS, helping to determine the satellite's orientation relative to Earth's magnetic field.

Nano-Satellite
A small satellite with a mass between 1 and 10 kilograms. Nano-satellites are often used for scientific research, technology demonstration, and educational missions due to their lower cost and ease of deployment compared to larger satellites.

Orbit Determination and Propagation
The process of determining a satellite's current orbit and predicting its future positions. This involves analyzing tracking data and applying mathematical models to calculate the satellite's trajectory, ensuring accurate mission planning and collision avoidance.

Public-Key Infrastructure (PKI)

A framework for secure communication and authentication using cryptographic keys. In satellite communications, PKI is used to ensure the integrity and confidentiality of data transmitted between satellites and ground stations.

Quantum Computing

A new computing paradigm based on the principles of quantum mechanics. Quantum computers have the potential to solve certain problems much faster than classical computers. In satellite technology, quantum computing could revolutionize data processing and encryption.

Quantum Key Distribution (QKD)

A method of secure communication that uses quantum mechanics to distribute encryption keys. QKD ensures that any attempt to intercept the keys can be detected, providing a high level of security for satellite communication links.

Reaction Wheel

A type of actuator used in satellite ADCS to control the satellite's orientation. Reaction wheels spin at varying speeds to generate torque, allowing precise control of the satellite's attitude without using propellant.

Remote Sensing

The acquisition of information about an object or area from a distance, typically using satellites equipped with sensors. Remote sensing is used for applications such as Earth observation, environmental monitoring, and disaster response.

Satellite Constellation

A group of satellites working together to provide global or regional coverage. Satellite constellations are used for applications such as communication, navigation, and Earth observation, offering continuous coverage and redundancy.

Software-Defined Radio (SDR)

A radio communication system where traditional hardware components are replaced with software. SDRs allow for dynamic reconfiguration of frequencies and communication protocols, enhancing the flexibility and adaptability of satellite communication systems.

Star Tracker

A device that uses the positions of stars to determine a satellite's orientation. Star trackers are highly accurate and are often used in conjunction with other sensors in the ADCS to maintain precise control of the satellite's attitude.

Telemetry
The process of collecting and transmitting data from a satellite to a ground station. Telemetry data includes information about the satellite's health, status, and environmental conditions, enabling ground operators to monitor and control the satellite.

Thermal Control System
A system used to maintain the temperature of satellite components within operational limits. Thermal control systems include passive elements like thermal blankets and radiators, as well as active elements like heaters and heat pipes, ensuring that the satellite operates efficiently in the extreme temperatures of space.

Wideband Global SATCOM (WGS)
A satellite communications system used by the U.S. Department of Defense. WGS satellites provide high-capacity communication links for military and government operations, utilizing advanced software-defined transponders for enhanced performance and flexibility.

14 LIST OF ACRONYMS

3D Three-Dimensional: Refers to objects or images having width, height, and depth.

ADCS Attitude Determination and Control System: A system used in satellites to control and stabilize their orientation in space.

AES Advanced Encryption Standard: A symmetric encryption algorithm used to secure data.

AI Artificial Intelligence: The simulation of human intelligence in machines that are programmed to think and learn.

ASIC Application-Specific Integrated Circuit: A custom-designed integrated circuit for a particular use, rather than for general-purpose use.

CCSDS Consultative Committee for Space Data Systems: An international organization that develops standards for space data and information systems.

DARPA Defense Advanced Research Projects Agency: An agency of the U.S. Department of Defense responsible for the development of emerging technologies for use by the military.

DHU Data Handling Unit: A component in satellites responsible for managing the flow of data from the payload to the communication system.

DSP Digital Signal Processing: The use of digital processing techniques to analyze, modify, or synthesize signals.

ECSS European Cooperation for Space Standardization: An organization that develops standards for space projects within Europe.

EOS Earth Observing System: A program by NASA for long-term global observations of the land surface, biosphere, atmosphere, and oceans of the Earth.

ESA European Space Agency: An intergovernmental organization dedicated to the exploration of space.

FGPA Field-Programmable Gate Array: An integrated circuit that can be configured by the user after manufacturing to perform various tasks.

GPS Global Positioning System: A satellite-based navigation system that provides location and time information anywhere on or near the Earth.

HIL Hardware-in-the-Loop: A technique used for the development and testing of complex real-time embedded systems, where physical parts of the system are replaced with virtual simulations.

HPC High-Performance Computing: The use of supercomputers and parallel processing techniques for solving complex computational problems.

IDS Intrusion Detection System: A device or software application that monitors network or system activities for malicious activities or policy violations.

IoT Internet of Things: The network of physical objects embedded with sensors, software, and other technologies to connect and exchange data with other devices and systems over the internet.

ISS International Space Station: A space station, or habitable artificial satellite, in low Earth orbit.

ITU International Telecommunication Union: A specialized agency of the United Nations responsible for issues that concern information and communication technologies.

JAXA Japan Aerospace Exploration Agency: The Japanese national aerospace agency responsible for space research and exploration.

LEO Low Earth Orbit: An orbit around Earth with an altitude between 160 kilometers (99 mi) and 2,000 kilometers (1,200 mi).

ML Machine Learning: A branch of artificial intelligence that involves the use of data and algorithms to imitate the way that humans learn, gradually improving its accuracy.

MODIS Moderate Resolution Imaging Spectroradiometer: An instrument aboard NASA's Terra and Aqua satellites that collects data in 36 spectral bands for monitoring global dynamics.

NASA National Aeronautics and Space Administration: An independent agency of the U.S. federal government responsible for the civilian space program, as well as aeronautics and aerospace research.

NATO North Atlantic Treaty Organization: An intergovernmental military alliance between 30 North American and European countries.

PKI Public-Key Infrastructure: A framework for creating a secure method for exchanging information based on public key cryptography.

QKD Quantum Key Distribution: A secure communication method that uses quantum mechanics to enable two parties to produce a shared random secret key.

RF Radio Frequency: The frequency range used for radio communication.

SDR Software-Defined Radio: A radio communication system where components that have typically been implemented in hardware are instead implemented by means of software.

SIL Software-in-the-Loop: A method of testing

and validating software for embedded systems using a simulation environment.

UHF Ultra High Frequency: The radio frequency range between 300 MHz and 3 GHz, used for television broadcasting, mobile phones, satellite communication, and other applications.

VHF Very High Frequency: The radio frequency range between 30 MHz and 300 MHz, used for FM radio, television broadcasts, and other communication applications.

WGS Wideband Global SATCOM: A satellite communications system used by the U.S. Department of Defense to provide high-capacity communication links.

ZTD Zenith Total Delay: The total delay of a signal caused by the atmosphere, measured in the zenith direction, affecting satellite communication and positioning accuracy.

Beyond Hardware: An Introduction to Software-Defined Satellites

15 TECHNICAL SPECIFICATIONS

Understanding the technical specifications of software-defined satellites is crucial for appreciating the advancements and capabilities that these technologies bring. This chapter provides a comprehensive overview of the key technical specifications associated with software-defined satellites, covering areas such as hardware components, software systems, communication protocols, and performance metrics.

Hardware Components

1. Field-Programmable Gate Arrays (FPGAs)
   Description: Reconfigurable integrated circuits that can be programmed after manufacturing to perform various tasks.

Specifications:

Logic Cells: Typically range from 10,000 to over 1 million logic cells.

Clock Speed: Between 100 MHz to 1 GHz.

Memory: Embedded RAM blocks ranging from a few KB to several MB.

I/O Pins: Up to 1,000 input/output pins.

2. Onboard Computers (OBC)

Description: Central processing units that manage the satellite's operations, data processing, and communication.

Specifications:

Processor Type: Radiation-hardened processors like the LEON3 or PowerPC.

Clock Speed: Typically 200 MHz to 1 GHz.

Memory: RAM ranging from 256 MB to 4 GB; non-volatile storage from 4 GB to 128 GB.

Operating System: Real-time operating systems (RTOS) such as VxWorks or RTEMS.

3. Communication Systems

Description: Systems that manage the transmission and reception of data between the satellite and ground stations.

Specifications:

Software-Defined Radios (SDRs):

Frequency Range: Typically 30 MHz to 6 GHz.

Bandwidth: Up to 200 MHz.

Modulation Schemes: BPSK, QPSK, 8PSK, 16QAM, etc.

Antenna Systems:

Type: Patch antennas, helical antennas, phased array antennas.

Gain: 6 dBi to 30 dBi, depending on the antenna type and frequency.

4. Power Systems

Description: Components that generate, store, and distribute electrical power to the satellite's subsystems.

Specifications:

Solar Panels:

Efficiency: 28% to 32%.

Power Output: Typically 100 W to 3 kW, depending on size and technology.

Batteries:

Type: Lithium-ion or lithium-polymer.

Capacity: 20 Ah to 200 Ah.

Voltage: 28 V nominal.

5. Thermal Control Systems

Description: Systems that manage the satellite's temperature to ensure all components operate within safe limits.

Specifications:

Passive Components: Multilayer insulation (MLI), thermal coatings.

Active Components: Heaters, heat pipes,

thermal louvers.

Temperature Range: Operating temperature typically between -20°C to +50°C.

Software Systems

1. Operating Systems

Description: Software platforms that manage hardware resources and provide services for satellite applications.

Specifications:

RTOS: VxWorks, RTEMS, or custom real-time operating systems.

Kernel Size: Typically a few MB.

Latency: Low-latency operation, typically in the microseconds range.

2. AI and Machine Learning Algorithms

Description: Algorithms used for autonomous operations, data processing, and decision-making.

Specifications:

Types: Supervised learning, unsupervised learning, reinforcement learning.

Frameworks: TensorFlow, PyTorch, custom frameworks.

Processing Power: Utilizes onboard GPUs or specialized AI accelerators.

3. Data Handling and Compression

Description: Software systems for managing, processing, and compressing data collected by the satellite's sensors.

Specifications:

Data Rates: Typically 1 Mbps to 1 Gbps, depending on the payload.

Compression Algorithms: Lossless (e.g., PNG, ZIP) and lossy (e.g., JPEG, MPEG).

Storage Management: Dynamic allocation of storage resources based on mission needs.

4. Security Systems

Description: Software systems for ensuring the security and integrity of satellite operations and data.

Specifications:

Encryption: AES-256 for data at rest and in transit.

Authentication: Public-key infrastructure (PKI) for secure access.

Intrusion Detection: Real-time monitoring and anomaly detection algorithms.

Communication Protocols

1. Telemetry, Tracking, and Command (TT&C)

Description: Protocols for monitoring the satellite's health and status, and for sending commands from ground control.

Specifications:

Frequency Bands: S-band (2-4 GHz), X-band (8-12 GHz).

Data Rates: 100 kbps to 10 Mbps.

Protocols: CCSDS Space Packet Protocol, TCP/IP for higher layers.

2. Data Transmission

Description: Protocols for transmitting scientific and payload data from the satellite to ground stations.

Specifications:

Frequency Bands: Ka-band (26.5-40 GHz), Ku-band (12-18 GHz).

Data Rates: Up to 1 Gbps.

Modulation Schemes: QPSK, 8PSK, 16QAM.

3. Inter-Satellite Links

Description: Communication links between satellites in a constellation for data relay and network management.

Specifications:

Frequency Bands: Ka-band, V-band (40-75 GHz).

Data Rates: 10 Mbps to 1 Gbps.

Protocols: Proprietary or standardized protocols like CCSDS Proximity-1.

Performance Metrics

1. Data Throughput

Description: The amount of data a satellite can transmit to the ground stations within a given time period.

Specifications:

Measurement: Typically measured in Mbps or Gbps.

Typical Values: 100 Mbps to 1 Gbps, depending on the satellite's communication system.

2. Latency

Description: The time delay between sending a command to the satellite and receiving a response.

Specifications:

Measurement: Typically measured in milliseconds (ms).

Typical Values: 10 ms to 500 ms, depending on the distance and communication system.

3. Reliability

Description: The ability of the satellite to perform its intended functions without failure over its operational life.

Specifications:

MTBF (Mean Time Between Failures): Typically measured in years.

Typical Values: 5 to 15 years, depending on the satellite's design and mission requirements.

4. Power Efficiency

Description: The efficiency with which the satellite converts and uses electrical power.

Specifications:

Measurement: Power consumption in watts (W) and efficiency percentage.

Typical Values: Solar panel efficiency of 28% to 32%, overall power consumption of 500 W to 3 kW.

5. Thermal Performance

Description: The ability of the satellite's thermal control system to maintain safe operating temperatures.

Specifications:

Temperature Range: Operating temperature range for different components.

Typical Values: -20°C to +50°C for most electronic components.

This detailed overview of technical specifications provides a comprehensive understanding of the components, systems, and performance metrics that define software-defined satellites. By examining these specifications, we gain insight into the complexity and sophistication of these advanced space systems, as well as the technological innovations that drive their development and operation.

Understanding these technical details is essential for appreciating the capabilities and challenges associated with software-defined satellites. As the technology continues to evolve, these specifications will serve as benchmarks for future advancements, guiding the development of next-generation satellites that are even more flexible, efficient, and capable of meeting the dynamic needs of space missions and applications.

16 REFERENCES AND FURTHER READING

Books

1. "Satellite Communications Systems: Systems, Techniques and Technology"
   Authors: Gerard Maral, Michel Bousquet
   Publisher: Wiley
   Summary: This comprehensive book covers the fundamentals of satellite communications, including system design, modulation techniques, and the latest advancements in software-defined technologies.

2. "Artificial Intelligence and Machine Learning for Multi-Domain Operations Applications"
   Authors: S. Sitharama Iyengar, N. Balakrishnan
   Publisher: Springer

Summary: The book delves into the use of AI and machine learning in various domains, including space operations, providing detailed case studies and technical insights.

3. "Quantum Communication and Information Technology"
Authors: Alexander Sergienko, Shun Lien Chuang
Publisher: Springer
Summary: This book explores the principles of quantum communication and information technology, focusing on their applications in secure satellite communications and advanced data processing.

4. "Spacecraft Thermal Control Handbook: Fundamental Technologies"
Authors: David G. Gilmore
Publisher: The Aerospace Corporation
Summary: A detailed guide to thermal control technologies for spacecraft, covering both passive and active systems, and the role of software in optimizing thermal management.

5. "Field-Programmable Gate Arrays (FPGAs): Principles and Practices"
Authors: Clive Maxfield
Publisher: Newnes
Summary: This book provides an in-depth

understanding of FPGAs, their design principles, and practical applications in various fields, including space technology.

Online Resources and Journals

1. IEEE Xplore Digital Library
Summary: A comprehensive resource for accessing research papers, technical articles, and conference proceedings related to electrical engineering, electronics, and computer science, including topics on software-defined satellites and space technology.

2. Journal of Spacecraft and Rockets
Summary: Published by the American Institute of Aeronautics and Astronautics (AIAA), this journal features peer-reviewed papers on the design, development, and application of spacecraft and rockets.

3. NASA Technical Reports Server (NTRS)
Summary: An online database providing access to NASA's current and historical technical literature, including research papers, technical reports, and conference papers on satellite technology and space missions.

4. ESA's Open Access Repository
Summary: A platform for accessing ESA's research publications, technical documents, and

project reports, covering a wide range of topics related to space science and technology.

5. Defense Technical Information Center (DTIC)

Summary: A repository of technical reports and research documents from the U.S. Department of Defense, including studies on satellite communications, cybersecurity, and quantum technologies.

Beyond Hardware: An Introduction to Software-Defined Satellites

Andrew P. Whittaker

FURTHER BOOKS BY THE AUTHOR:

"An Introduction to Robustness in Satellite Operations" Lambert Academic Publishing, ISBN 978-3-659-87662-2

Andrew P. Whittaker

Beyond Hardware: An Introduction to Software-Defined Satellites

Andrew P. Whittaker

www.ingramcontent.com/pod-product-compliance
Lightning Source LLC
Chambersburg PA
CBHW071512220526
45472CB00003B/990